SpringerBriefs in Applied Sciences and Technology

Volume 7

T0183938

Jacek Korec

Low Voltage Power MOSFETs

Design, Performance and Applications

 Springer

Jacek Korec
Texas Instruments
Power Stage BU
116 Research Drive
Bethlehem, PA 18015, USA
korec@ti.com

ISSN 2191-530X   e-ISSN 2191-5318
ISBN 978-1-4419-9319-9   e-ISBN 978-1-4419-9320-5
DOI 10.1007/978-1-4419-9320-5
Springer New York Dordrecht Heidelberg London

Library of Congress Control Number: 2011924343

Printed on acid-free paper

Springer is part of Springer Science+Business Media (www.springer.com)

Introduction

This is not intended to be a reference book. This is not intended to be a student book. This script should help newcomers to the power management community like young device and circuit designers, product engineers, and marketing staff entering into the field of low voltage power MOSFETs and their applications.

The goal is to show the relations between the way how a power MOSFET has been designed and the performance of this device in different applications. No power device expertise is required from the reader. Just a general electric engineering background should be enough to find fun in understanding why some MOSFETs are better than other, and how to use a better MOSFET performance to improve power management products.

An ideal MOSFET should have zero conduction resistance and zero switching losses. This is trivial. How to approach the ideal is not trivial as demonstrated by the many MOSFET generations introduced into the market since early 1980s. Basically, a power MOSFET is a piece of Silicon integrating a large number of basic cells acting in parallel. It has three terminals: Source and Drain being main power connectors, and a Gate used as the control to switch the current. Usually, the Silicon chip is molded into a plastic package which has exposed leads acting as the mentioned terminals. The performance of the package is as important as the performance of the Silicon device itself. Package should add minimum on serial resistance, should have minimal parasitic inductance, and a low thermal impedance allowing easy dissipation of the generated heat. In spite of the importance of package characteristics, this script will focus on the design of the Silicon device itself and package performance will be treated marginally.

Because of the restricted scope and character of the presented material, no exact mathematic formulas will be used to explain the discussed phenomena. Instead of exact but complex theories, the issues treated here are explained qualitatively in a plain language using schematic illustrations, easy to follow. Any details can be rather found in reference books, if only the whole picture is clear and properly understood.

Contents

Chapter 1
MOSFET Basics

There is 90% chance that the reader can skip the 1st chapter as it covers just the basics of MOSFET structure and operation, along some examples of typical applications. Possibly the best approach is to have a glance at the figures and skip the text if the enclosed illustrations are well understood.

1.1 Structure and Operation

Basic MOSFET structure is shown in Fig. 1.1 for the case of an N_{ch} transistor. In general this script is focusing on N_{ch} devices, and the few P_{ch} cases are clearly stated.

In the IC technology, MOSFETs are usually formed inside of a P-well which also provides the body region of the transistor. In the case of discrete power MOSFETs, body is formed by dedicated implantations before or after gate deposition and patterning. If body and source/drain are implanted after gate formation, all implants are aligned to the edges of the gate, and body definition is not sensitive to possible misalignment of the gate structure. Such process is called Double Diffused MOSFET (DMOS). MOSFET operation is based on the control of an inversion layer (channel) built underneath of the gate, along the body/oxide interface. The related impact on the band gap structure of the MOS stack is illustrated by Fig. 1.2.

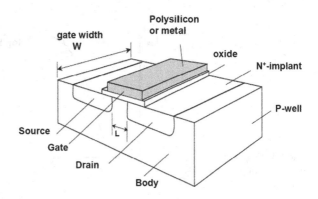

Fig. 1.1 Schematic illustration of a basic N_{ch} MOSFET structure

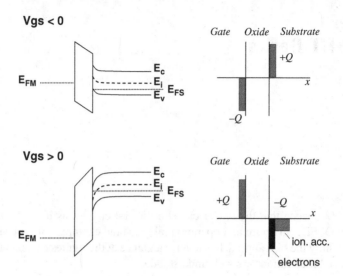

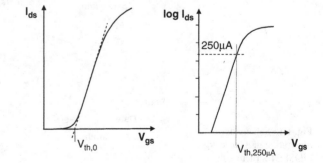

Fig. 1.2 Illustration of the MOS band gap structure under applied gate potential

As the body region has a P-type doping, applying a negative bias to the gate results in building a positive charge along the silicon/oxide interface by accumulation of holes. In the opposite case, applying a positive bias to the gate drives the holes away from the MOS interface leaving a depleted region with a negative charge of ionized acceptor atoms. If the gate bias is strong enough, the bands are bent further, and eventually Fermi level (E_{FS}) approaches conduction band (E_c). This leads to generation of electrons along the MOS interface by injection from neighbor PN junctions or by thermal generation. Electrons are mobile carriers building a conductive channel underneath of the gate. This condition is called strong inversion, and the gate bias at the onset of the strong inversion is called threshold voltage (V_{th}). The value of the threshold voltage can be calculated using equations describing the bending of the band diagram, and the extracted value is usually provided by FET simulation tools. However, everyday's approach has to be simple, so two engineering definitions have been implemented (Fig. 1.3).

Fig. 1.3 Engineering definitions of the threshold voltage

A linear threshold voltage can be extracted from the transfer characteristics ($I_{ds}-V_{gs}$) by reading the intercept of the extrapolation of the linear portion of the curve ($V_{th,0}$). This value is close enough to the physical definition of V_{th} by strong inversion condition, and is widely used by device engineers. The second definition is based on reading the gate voltage corresponding to a pre-defined transistor I_{ds} current (e.g. 250 μA) from the sub-threshold curve ($V_{th,250\,\mu A}$). This V_{th} definition is popular by marketing and procurement people, although the corresponding value is lower than $V_{th,0}$ by 0.2–0.3 V. The other issue with $V_{th,250\,\mu A}$ is that its value depends on the die size, so different MOSFET products manufactured with the same technology will have different $V_{th,250\,\mu A}$ values.

Talking about applications of MOSFETs, two operation modes are distinguished (Fig. 1.4):

– linear region, where $V_{ds} \ll V_{gs} - V_{th}$ leading to $I_{ds} \sim (V_{gs} - V_{th})V_{ds}$
– saturation region, where $V_{ds} > V_{gs} - V_{th}$ leading to $I_{ds} \sim (V_{gs} - V_{th})^2$

In both regions the transconductance factor g_m is affected by the MOSFET design:

$$g_m = \Delta I_{ds}/\Delta V_{gs} \sim C_{ox}{}^* W/L$$

which means, the thinner the gate oxide ($C_{ox} = \varepsilon_0/t_{ox}$) and shorter the channel (L), the larger I_{ds} for the same V_{gs} and V_{ds}.

A power MOS transistor usually operates as a switch along a load characteristics represented in Fig. 1.5 by a dashed line. Here, the turn-on and turn-off operating points (A and B in Fig. 1.5) follow the signal applied to the gate terminal (V_{gs}).

The I-V waveforms across a MOSFET switching a resistive load are illustrated by Fig. 1.6.

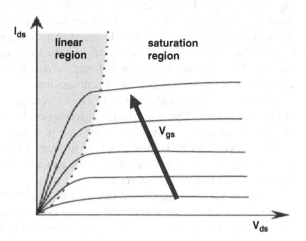

Fig. 1.4 Output $I_{ds} - V_{ds}$ characteristics showing linear and saturation mode

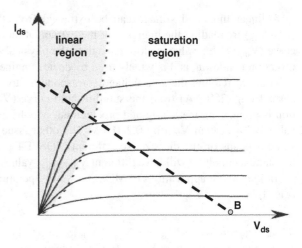

Fig. 1.5 Output $I_{ds} - V_{ds}$ characteristics showing switching operating points

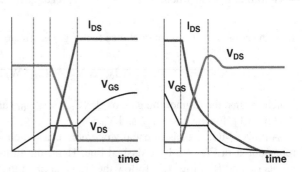

Fig. 1.6 Schematic switching waveforms for resistive load

During turn-on, drain current starts to flow as the gate voltage exceeds V_{th} value. A collapse of the drain voltage is associated with discharging of the so called Miller capacitance (C_{gd}). During turn-off, the device goes through a reverse sequence of events. First, blocking voltage is build across the switch as a depletion region is developed inside of the FET. This is observed as a V_{gs} plateau during charging of C_{gd}. As the gate bias falls down below V_{th}, the drain current decays to zero. The overlap of current and voltage waveforms across the switch produce power loss which has to be dissipated in form of heat removal.

In both cases, smaller C_{gd} results in a faster voltage transient – so in other words in faster switching. The amount of the overlap of the I-V waveforms depends on the type of the load. For example in the case of a mixed resistive/capacitive load, the current will lead the voltage, resulting in a larger turn-on loss, and smaller turn-off loss. In any case, faster switching leads to a smaller area of the overlap of the waveforms, i.e. in less total power loss.

Finally it has to be mentioned that four types of MOSFETs are distinguished. If source and drain have N-type doping, and body region is doped P-type, we are dealing with an N_{ch} device (see Fig. 1.7). Source and body regions have a common

Fig. 1.7 Enhancement and
depletion N$_{ch}$ MOSFET
symbols

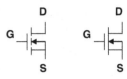

Fig. 1.8 Enhancement and
depletion P$_{ch}$ MOSFET
symbols

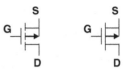

contact, so that V$_{gs}$ bias is applied relative to the body potential. The drain bias is typically positive, and a positive bias has to be applied to the gate to turn the switch on. If V$_{th}$ has a positive value, the switch is turned-off at zero gate bias, and the MOSFET is called an enhancement MOSFET or a normally-off switch. In the case of a negative V$_{th}$ value, the device is already on at zero gate bias. Such a device is called a depletion MOSFET or a normally-on switch. If a reverse bias is applied to drain, the internal body/drain diode conducts in parallel to the channel conduction which is still controlled by the gate potential.

A P$_{ch}$ MOSFET can be created by reversing the polarity of the doping of source/drain and body regions. Now normal operation requires a negative drain and gate bias, and the source/body potential is tied to the hot rail. Also here, enhancement and depletion devices are distinguished according to the normally-off or normally-on state at zero gate bias (see Fig. 1.8).

1.2 Application Examples

MOSFETs have been initially introduced in IC design as current switches at a given drain bias. Thus, in digital ICs MOSFETs represent simple on and off states. In analog ICs they are used as simple current breakers and/or current amplifiers. The very basic example of a digital CMOS inverter circuit is shown in Fig. 1.9. Typically, in IC technology body region of a group of transistors is formed in a well attached to the high rail (P$_{ch}$) or ground potential (N$_{ch}$). Individual MOSFETs have separate source and drain contacts.

A CMOS inverter transforms a low level gate signal into a stronger clock signal with opposite polarity. Popular schemes of small signal and large signal amplifiers used in analog ICs are shown in Figs. 1.10 and 1.11.

On the other hand, MOSFETs used in power management solutions ("power MOSFETs") are just switching elements which are supposed to turn on and off current against an applied voltage with minimum conduction and switching losses. Typical power applications are illustrated by Figs. 1.12, 1.13 and 1.14.

Fig. 1.9 CMOS inverter

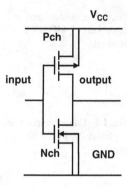

Fig. 1.10 Principle of small
signal amplification

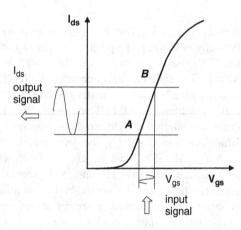

Fig. 1.11 Principle of large
signal amplification

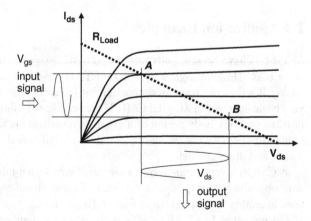

Fig. 1.12 Load switch on
low side position

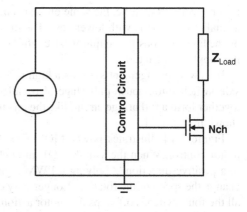

Fig. 1.13 Load switch on
high side position

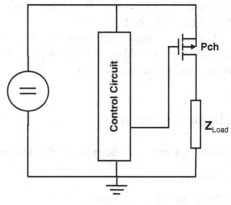

Fig. 1.14 Schematics of a
motor control circuit

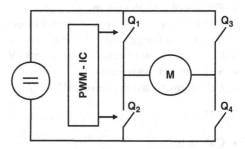

Using an N_{ch} MOSFET as a load switch it is easier to place it on the low side position as gate signal can be ground related. The opposite holds for P_{ch} MOSFETs which operate well on the high side position. High side switches are popular in applications like automotive where the loads have to be grounded when turned-off.

Commonly, power applications require normally-off switches, as the loads have to be disconnected during start-up sequence of the system. For low current applications in mA load range, P_{ch} MOSFETs can be used. In the case of large switched currents N_{ch} MOSFETs are usually preferred as the higher electron mobility offers lower specific $R_{ds,on}$, i.e. the same on-resistance of the switch can be achieved with a smaller die, that is with lower cost. The drawback is that using an N_{ch} MOSFET on the high side position requires the control circuit to supply a gate bias higher than the supply voltage rail.

In power management systems a high side switch is often combined with a low side switch into a push-pull stage also called a half-bridge. Two such pairs put together form a full bridge circuit like the one used in the example of a motor control shown in Fig. 1.14.

Figure 1.14 illustrates power MOSFETs application for a variable speed DC motor control. When the switches Q1 and Q4 are closed the DC motor is turning in a positive direction. Applying a PWM signal to these switches it is possible to change the speed of the rotation. Longer duty cycle leads to a higher speed. Turning all the four switches off stops the motor action. Closing Q3 and Q2 switches results in a reverse motor action. Here, once again, the rotation speed is adjusted by the duty cycle controlled by the IC.

The most popular application of power MOSFETs in power management solutions are Switched Mode Power Supplies (SMPS). SMPS systems are discussed in some detail in the next chapter.

Chapter Summary

Following topics have been handled in this chapter:

Section 1.1:

* Basic MOSFET structure
* Physics of a MOS system
* MOSFET operation by gate control of the channel
* Physical and engineering definitions of threshold voltage
* Linear and saturation regions in output IV characteristics
* Switching IV waveforms
* Enhancement and depletion MOSFETs

Section 1.2:

* CMOS inverter
* Small and large signal amplification
* Load switching and motor control applications

Chapter 2
Application Requirements

The material presented in this script covers low voltage applications extending from battery operated portable electronics, through POL-converters (Point of Load), internet infrastructure, automotive applications, to PC's and server computers. Thus, the switched current can be as low as hundreds of mA, or as high as 30 A per switch. The switched voltage can vary from 3 V for battery supply to 12 or 24 V as an intermediate voltage in power distributed systems. Even though the switched power varies respectively from 1 W to 1 kW, the application related issues have many commonalities. Just to stay focused, medium to high voltage applications have been excluded from the discussion.

Traditionally, system engineers require MOSFET designed for a maximum drain voltage ($V_{ds,max}$) higher by 20% than the maximum voltage spike value observed across the MOSFET switch during circuit operation. Following this rule, 5 V MOSFETs can be used for 3 V applications, 12 V circuits require MOSFETs designed for $V_{ds,max}$ of 25 or 30 V, and 24 V applications use 40–60 V MOSFETs.

Next basic requirement defines maximum allowed $R_{ds,on}$ of the switch and is derived from thermal considerations. Conducting current produces power loss (I^2*R) generating heat which has to be dissipated. The generated heat flows from the power source which is within the switching device (Si die), through the MOSFET package to the heat sink, or just to the PCB (Printed Circuit Board). This procedure imposes three thermal limits:

- the temperature of the Si die can not exceed $T_{j,max}$ (maximum junction temperature) specified in the MOSFET data sheet (usually 150°C),
- the dissipated power can not heat PCB to a temperature higher than allowed by the PCB manufacturer (usually 100° to 110°C),
- the total heat produced by the electronic system can not exceed some thermal limit defined for the equipment. For example server computers in data processing centers can not produce more heat than a maximum value allowed per floor square foot – otherwise A/C costs become prohibitive.

The situation is even worse if the switch is permanently switching current at a high frequency. Switching power loss produced by the MOSFET is defined by the overlap

J. Korec, *Low Voltage Power MOSFETs*, SpringerBriefs in Applied Sciences and Technology 7, DOI 10.1007/978-1-4419-9320-5_2, © Jacek Korec 2011

of switched current and voltage waveforms. Additionally, total power loss has to include power loss generated by the MOSFET driver ($Q_g * V_{gs} * f$).

Each application has an optimum sized MOSFET. If the die is too small, MOSFET $R_{ds,on}$ and the resulting conduction loss is too large. If the die is too large, MOSFET associated capacitances are too large, and so is the resulting switching power loss.

Finally, the most important 3 requirements imposed by all commodity applications are cost, cost, and once again cost.

2.1 Load Switching

One can say load switching is all about cost. In principle yes, but a good technical solution has to consider thermal limitations, some circuits have to be placed within limited physical space, and in some applications MOSFETs have to be switched with very small gate charge injected by the driver circuit.

When looking for a solution within given cost and thermal limits, one has to deal with the $R_{ds,on}$ of the switch, i.e. the cost of Si die, and the cost and the thermal impedance of the package. On-resistance of the switch is defined by specific on-resistance given by the technology used and active area of the Silicon chip (R_{sp} / A) plus the contribution of the package resistance. Silicon die cost is defined by the type of the technology used (wafer size and cost) and the die area. Thermal performance, size and cost of the package depend on the type of the package used.

Talking about low voltage power management applications the choice of the MOSFET switch is often dictated by maximum load current to be switched.

Power IC's often integrate lateral MOSFETs exhibiting on-resistance in the range of 100 mΩ. If the on-resistance has to be much lower, a less expensive solution is to restrict the IC part to the controller, and use external MOSFETs. A preferred approach in this case is a multi-chip module (MCM), where the same package contains IC and MOSFET chips. The reason for it is qualitatively explained in Fig. 2.1.

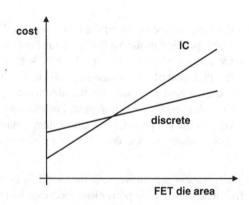

Fig. 2.1 Integrated FET cost
vs. discrete assembly

For sake of this discussion let us assume that the MOSFET device can be manufactured with 8 mask levels, and the IC circuit needs 24 mask process which leads
to a triple cost per wafer. So, even with higher assembly cost of the MCM solution,
the multi-chip approach is more cost effective for FETs consuming large die area.

In many cases, monolithic integration of the power stage is too expensive if the
power stage area approaches 30% of the total IC. For most combinations of IC /
MOSFET technologies, a more cost effective solution can be achieved by using
external FETs if the required current capability exceeds 3–5 A. Monolithic integration of FETs for 5–15 A applications is technically feasible when thick Cu
metallization is used, but this approach is seldom economical.

Integration of multiple independent switches on one die requires electrical isola-
tion of individual devices. This is achieved by placing individual FETs in isolated
PN wells, and requires a lateral design of the switches. This method, although popular in power ICs, puts a restriction on scaling integrated MOSFETs to large areas.
Power MOSFETs are built by parallel connection of a large number of small cells,
with a small pitch ranging a few microns. Figure 2.2 illustrates this situation for a
stripe design of the basic MOSFET cells.

Fig. 2.2 Drain and source
bus structure on top of lateral
MOSFETs

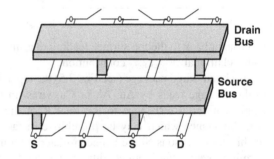

Source and drain contacts are connected alternatively to source and drain buses
(1st metal layer), which are then connected to large area terminal pads made in an
upper metal layer (2nd metal layer) not shown in the picture. The issue is that flow
of current through such a bus structure results in a voltage drop along metal lines
leading to gate voltage de-biasing, which in turn increases $R_{ds,on}$ of the switch.

A natural solution of this problem is achieved in discrete MOSFETs by placing
one power electrode on the top surface, and the other electrode on the back side of
the die, as illustrated in Fig. 2.3 for the case of a drain down technology.

Three package types have been mentioned in Table 2.1: CSP, wired, and clip
package. CSP stands for Chip Scale Package and is actually a package without
package.

In a CSP device all terminals are placed on the top surface of the wafer in form of
solder bumps, and the only encapsulation of the product is provided by the passiva-
tion film covering the patterned top metal layer (see Fig. 2.4). Here, this is the only
situation where lateral devices can have an advantage over vertical device structure.
In the case of a device with a vertical current flow the substrate connection has to

Fig. 2.3 Placement of power terminals on a FET with vertical current flow

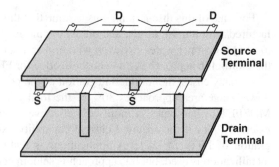

Table 2.1 MOSFET choice dictated by designed power range

I_{max} (A)	$R_{ds,on}$ (mΩ)	Assembly type	$R_{package}$ (mΩ)
<3	30–200	Power IC, CSP, wired package	5–10
3–10	10–30	Power IC, wired package	2–5
10–20	3–10	Wired or clip package	0.3–2
>20	1–3	Clip package	0.2–0.3

be brought back to the top surface leading to a small "dead" area on the Si chip and some additional "$R_{package}$" contribution.

In a bond-wired package (see Fig. 2.5) the top terminal of the device is connected to the package leads by Au, Al, or Cu wires. In each case bond-wires introduce significant serial resistance to the total R of the switch, and create some parasitic inductance which may be prohibitive in fast switching applications. Mostly for historical reasons Si die is encapsulated by a molded plastic body. Nowadays, with improved clean room conditions in assembly line, the plastic molding can be skipped for wireless packages.

In a clip package the top electrode is created by a Cu strap connection to the lead post (see Fig. 2.5). The introduction of clip packages in late 90' presented a big improvement in the package performance, as $R_{package}$, thermal impedance, and parasitic inductance have been minimized. As usual, technical improvement means that the better package is more expensive, as the required solderable top metal increases wafer cost. For this reason, old fashioned wired packages are still in use

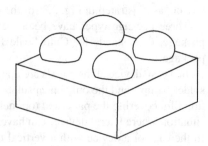

Fig. 2.4 CSP (chip scale package) device

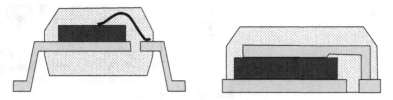

Fig. 2.5 Schematic drawing of wired and clip packages

for low power applications where low product cost is most important. On the other hand, high power applications can benefit from an enhanced thermal performance of a modified clip package, where the heat can be dissipated also through the top of the package surface.

Placing a heat sink on the top of the enhanced package can allow an increase of the maximum current density by a factor of two or more.

A special category of load switches are so called bypass switches which are formed by two back to back transistors as shown in Fig. 2.6.

Fig. 2.6 Common drain and common source bypass switches

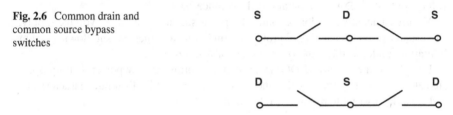

As we know from the basic operation principle, MOSFET can block current flow in one direction only. If a reverse drain bias is applied, the integral body diode of the MOSFET will conduct in parallel to the MOS channel as soon as V_{ds} drop exceeds about 0.6 V. If a bi-directional switch is required, two MOSFETs have to be used in a common drain or common source configuration. Bypass switches are often used in battery operated electronics to switch the circuitry between different battery packs and/or battery charger.

To turn-on a bypass switch formed by two P_{ch} transistors it is enough to pull their gates to ground. In the case of N_{ch} transistors a charge pump is usually used to generate a gate bias higher than the supply voltage attached to the hot rail. In spite of this complication, N_{ch} transistors are usually preferred than P_{ch} solution because of the lower $R_{ds,on}$ of the transistor for the same die size. As the transistor cost is proportional to the die size, N_{ch} solution means a lower cost for the customer.

2.2 Voltage Regulator Systems

Voltage regulator systems will be described using a battery operated electronic product as an example (Fig. 2.7). VRM stands form Voltage Regulator Module and this

Fig. 2.7 Schematics of a voltage regulator system

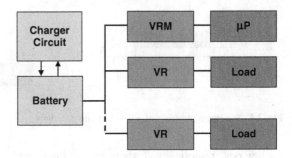

term is used for the power supply of a CPU (Central Processor Unit). Other VR blocks are used to stabilize the voltage supplied to individual load blocks in the whole system. The need for voltage regulation comes from the fact that the supply voltage from the battery vary between fully charged and discharged states, and the current demand of different load subsystems can change instantly. Thus, the task solved by VR blocks is input regulation and load regulation to assure a constant voltage output level within the allowed tolerance-band.

VR uses a negative feedback control loop to assure a stable output voltage supply, and usually provides additional protection functions like thermal shutdown, current limitation, under-voltage and over-voltage protection.

Low dropout regulator (LDO) is a popular solution for low power, battery operated electronics. It operates in the linear mode of the MOSFET output characteristics and acts like a variable resistor, see Fig. 2.8.

Fig. 2.8 LDO circuit scheme

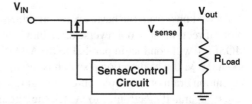

LDO efficiency is good as long as the output voltage level is close to the input voltage value (>80%). Otherwise the voltage difference multiplied by the load current creates a power loss which has to be dissipated in form of heat. So, the main requirement on a MOSFET used as an LDO is effective heat dissipation. Here,

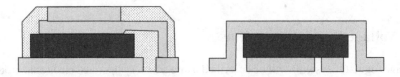

Fig. 2.9 Thermally enhanced clip packages

thermally enhanced packages like shown in Fig. 2.9 in conjunction with a heat sink placed on the top are a good solution.

In case of larger power consumption and/or a large voltage step between the input and the output voltage, a switched DC/DC converter offers better efficiency and is used instead of a linear regulator.

A DC/DC converter is build up by a PWM chopper (Pulse Width Modulation) and an LC output filter (Fig. 2.10). The PWM chopper includes a control IC, power switches, and a negative feedback loop monitoring the level of the output voltage and adjusting the duty cycle (D) of the chopper. Reacting to an increased current demand of the output load, control IC increases the duty cycle and supplies more energy to the output filter.

Fig. 2.10 Switched DC/DC converter

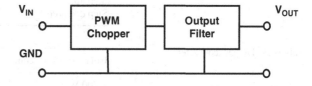

A higher switching frequency of the DC/DC converter enables a more precise control of the output voltage and allows the use of smaller components of the output filter, i.e. reduces the cost of the power supply. On the other hand, the increase of the switching frequency is limited by the allowed amount of heat generated by the power loss of the converter. Power loss is produced by conduction loss which is frequency independent and a switching loss which is proportional to the switching frequency. Total power loss is plotted in Fig. 2.11 as a function of the switching frequency for two MOSFET sets Q1 and Q2.

Fig. 2.11 Total power loss as function of switching frequency

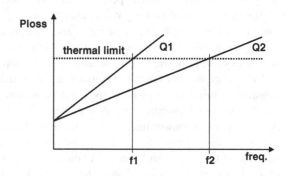

Transistors Q1 and Q2 have similar conduction losses, but transistor Q2 has significantly lower switching loss what allows an increase of the switching frequency of the converter from f1 to f2 before the system reaches the thermal limit. A breakdown of the power loss into individual contributions will be discussed in the chapter discussing MOSFET performance.

2.3 Switched Mode Power Supply

As stated before, a Switched Mode Power Supply (SMPS) offers much wider range of voltage regulation and better efficiency than a linear regulator. There are many topologies of SMPS systems. Here, a few examples will be mentioned (see Tables 2.2 and 2.3), and focus will be made on Synchronous Buck converter which is usually the power supply of choice for a variety of low voltage power management systems.

Table 2.2 DC/DC converter topologies

Non-isolated DC/DC converters	
Step down	Buck
Step up	Boost
Step down/up	Buck/boost

Table 2.3 DC/DC converter topologies

Isolated DC/DC converters	
< 200 W	Flyback Feed forward
200–400 W	Push-pull Half-bridge
> 400 W	Full bridge

The choice of SMPS topology is dictated by the power consumption level of the load. In the low power range (<100 W) non-isolated topologies are preferred.

Non-isolated DC/DC converters are PWM based hard switching circuits and all comments on switching converters made in the previous chapter apply to these topologies. There is a continuous demand on reduction of MOSFET switching loss in order to enable converter designers to increase the switching frequency, and by doing so to improve converter performance and reduce its cost.

The application of non-isolated converter topologies is limited to a rather small value of voltage step between the input and the output. The duty cycle can be reasonably controlled for voltage step values up to $10\times$. In the case of larger voltage step values and/or power levels, isolated converter topologies are used.

Isolated converters use a transformer to define the voltage step by the ratio of transformer turns on the primary (N_p) and the secondary side (N_s). An SPMS system for off-line applications needs a Power Factor Correction (PFC) block at the interface with the power line. In general the loads connected to the power line are not only resistive, but usually have some capacitive and/or inductive components which lead to a phase shift between voltage and current waveforms. The purpose of a PFC interface is to make sure that current waveform follows voltage, so that any power reflection into the power line is minimized. Figures 2.12, 2.13, and 2.14 show the structure of an off-line SMPS system, and examples of a PFC and converter blocks.

Isolated DC/DC converters are still using hard switching of the power switches resulting in power loss, and limiting the switching frequency. For high power applications (> 1 kW) resonant topologies are frequently used, as they adapt Zero Current

Fig. 2.12 Off-line SMPS system

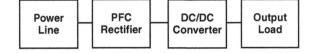

Fig. 2.13 Rectifier bridge with a PFC regulator

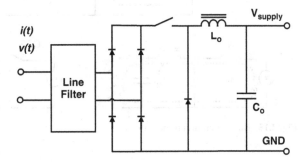

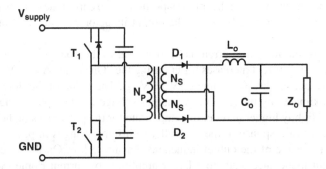

Fig. 2.14 Half-bridge isolated converter topology

(ZCS) or Zero Voltage (ZVS) switching schemes which reduce the stress and power loss of the switches. However, resonant topologies are more expensive and somehow restricted in the regulation range by the allowed range of frequency modulation.

Going back to low voltage applications, the operation of a Synchronous Buck converter (sync buck) will be discussed in some detail in the next chapter.

2.4 Synchronous Buck Operation

Basic operation of a buck converter is shown in Fig. 2.15. Controller IC defines the duty cycle of the PWM signal (D) based on the negative feedback signal monitoring the level of the output voltage. In the "On" phase, the power switch is closed and energy is stored in the output inductor. During the same time, the output inductor is supplying current to the output load and charging the output capacitor to the desired output voltage level.

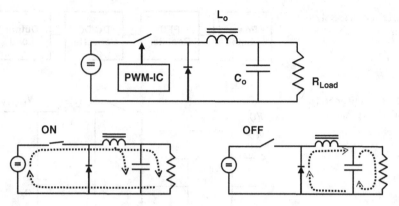

Fig. 2.15 Buck operation principle

During the "Off" time which is the rest of the switching period, the inductor current is decaying slowly, and the missing portion of the current demand is supplied from the capacitor. Figure 2.16 shows the output inductor current as a function of time.

The slope of the inductor current is proportional to the voltage drop across the inductor. Thus, the current increases faster during the "On" time ($\Delta V = V_{in} - V_{out}$) than it decays during the "Off" time ($\Delta V = V_{out}$). The value of the duty cycle D can be approximated by the ratio of the output voltage to the input voltage value, but has to be slightly higher in order to compensate for the power loss in the system.

To enable a fast response of the controller to instant changes in power demand of the load, the value of the output inductance has to be low. On the other side, a smaller output inductance leads to a larger amplitude of current ripple. So, a fast output regulation ability requires a high switching frequency keeping the switching period short. At the same time, high switching frequency allows a reduction of the output capacitance making the converter less expensive.

A basic implementation of a buck converter can be done using one control switch at the high side position, and a diode at the low side allowing for the free wheeling of the current during the "Off" time of the duty cycle. However, a large voltage drop

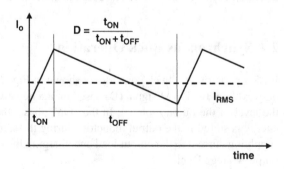

Fig. 2.16 Output inductor current

across the PN diode during conduction leads to considerable conduction power loss during the "Off" time, and lowers the efficiency of the converter. Efficiency of a converter system is defined by the ratio of the output power delivered to the load to the input power consumed from the power supply: the higher the power loss, the lower the efficiency.

The simple buck converter with a P_{ch} MOSFET at the high side position and a diode at the low side position is still in use for low output currents (< 1 A), where the cost is most important. Otherwise, a Synchronous Buck topology (Fig. 2.17) has been broadly introduced. Here, the diode is substituted by an N_{ch} MOSFET working in the 3rd Quadrant at a negative drain bias. If the MOSFET is turned-off, the current flows through the body diode. As soon as a positive bias is applied to the gate, the current flows mainly through the MOSFET channel.

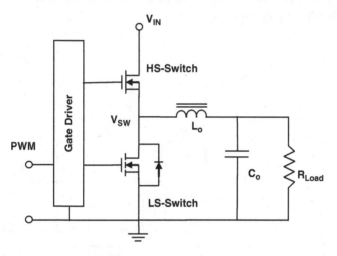

Fig. 2.17 Synchronous buck converter

A gate driver circuit implements the duty cycle defined by PWM clock signal, and is responsible for the right timing of the gate signals introducing a delay time between the operation of the HS and LS switches. This delay time called brake-before-make is needed to avoid a simultaneous turn-on of both transistors leading to a spontaneous cross current and destruction of the circuit.

The sequence of the events can be described following the illustration shown in Fig. 2.18:

– gate signal to the LS switch is turned-off,
– LS switch turns-off after a delay time depending on the size of the transistor (Td, off),
– as the LS gate potential goes below a pre-defined value, the HS gate signal is turned-on (the $t_{HS,delay}$ may be fixed or adjusted to the actual decay of $V_{gs,LS}$),
– HS switch turns-on after the inherent delay time of the transistor,

- LS body diode is conducting during the dead time when both MOSFETs are off (this can be observed as a period of negative bias of the LS drain),
- switch node potential (V_{SW}) shoots high as soon as the HS switch starts to conduct, and makes some oscillations due to the energy stored in the parasitic LC components of the power switches and PCB (Printed Circuit Board),
- during the high dv/dt swing of the switch node the LS gate potential bounces up which may result in a short shoot-through current (explained later),
- HS switch remains turned-on for the "On" time of the duty cycle,
- PWM signal turns the gate of the HS switch off,
- gate driver circuit implements a short delay time before the LS gate is turned-on,
- during the dead time, the body diode of the LS transistor conducts the output current,
- LS gate is turned-on to allow the LS MOSFET to conduct for the rest of the period time (free wheeling action).

Fig. 2.18 Gate driver timing in a sync buck converter

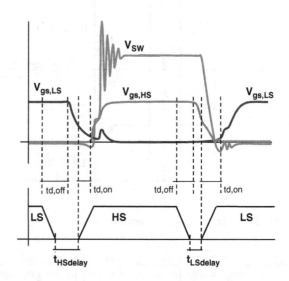

It can be noticed, that the delay time at the rising edge of the voltage at switch node has to be longer than the delay time at the falling edge of the switch node. For short duty cycles normally used in low voltage operation of sync buck converters, the LS switch conducts much longer time than the HS switch and the LS transistor die has to be made larger (low $R_{ds,on}$) than the HS. Larger die leads to a longer inherent delay time of the transistor (Td, off) when responding to the gate signal. Thus, HS switch responds faster than LS switch and there is a stronger need for a gate signal delay time at the rising edge than at the falling edge.

Of course, the delay times have to be made as short as possible to avoid excessive conduction time of the LS body diode. This can be done easier in the case of multi-chip modules (MCM) where the gate driver is integrated in the same package

Fig. 2.19 Illustration of the shoot-through effect

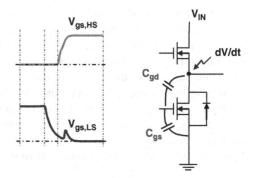

with the switching transistors, and the gate timing can be matched to the dynamic performance of the MOSFETs.

Other important issue in the operation of a sync buck converter is the danger of shoot-through. Shoot-through current may be induced by the feedback of the Miller effect creating a short rise of the gate bias, as shown in Fig. 2.19.

A high dV/dt at the switch node following the turn-on of the HS MOSFETs charges the Miller capacitance of the transistor (C_{gd}), this in turn injects charge into the gate node of the LS transistor. The injected charge has to sink down to ground through the output impedance of the gate driver, or will be absorbed by the C_{gs} of the LS switch. If the ratio of the C_{gd}/C_{gs} capacitance is larger than a critical value (depending on the sink resistance of the gate driver), the peak value of the voltage bounce at the LS gate may exceed the threshold voltage, and the LS transistor conducts for a short time. The shoot-through current flows from the voltage supply through the HS and LS transistors to the ground and results in a power loss which may be significant at light load condition. Also, the shoot-through current flows through the LS MOSFETs when the transistor exhibits a high drain bias. If such condition occurs repetitively, it may lead to reliability issues with the LS switch.

Schematic examples of sync buck converter efficiency for chip sets showing different performance are shown in Fig. 2.20.

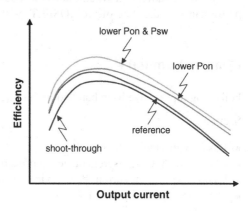

Fig. 2.20 Synchronous buck efficiency for different MOSFET chip sets

Chip set with smaller conduction losses show advantage in efficiency at heavy load condition. Smaller switching losses manifest themselves mostly around the peak efficiency, and additional power losses which are output current independent lower the efficiency at light load condition (shoot-through and gate driver loss).

Figure 2.21 shows the method of die size optimization for a chip set designed to work at known operating conditions (V_{in}, V_{out}, I_o, freq.). If the die area is too small, the resulting $R_{ds,on}$ of the switch is too large, and the respective conduction loss is too high. On the other hand, if the die area is too large, transistor capacitances are large leading to excessive switching loss. Both, conduction and switching power losses are balanced for the optimum size of the transistor die.

Fig. 2.21 Optimization of die sizes for HS and LS transistors

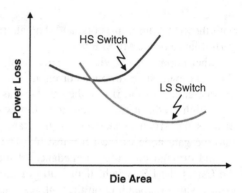

In a case of a large ratio of the input to the output voltage, for example when $V_{in} = 12$ V and $V_{out} = 1.2$ V, the duty cycle is small and LS transistor has to conduct much longer than the high side transistor. On the other hand, HS switch is switching against the full input voltage resulting in large switching loss. Thus, the LS switch has to be made much larger than the HS switch. In an opposite case, when the input and the output voltage are comparable (e.g. for a ratio of 2:1), both MOSFETs can be made equally large.

The optimization of a synchronous buck design using the advantage of improved performance of the state-of-art MOSFETs will be discussed in the last chapter.

Chapter Summary

Following topics have been handled in this chapter:

Section 2.1:

- Basic requirements defining $R_{ds,on}$
- Integrated FET vs. discrete device approach
- Lateral vs. vertical current flow devices
- Bond-wired and clip packages

Section 2.2:

- VRM systems
- LDO regulator
- Switched DC/DC converter
- Conduction and switching power loss

Section 2.3:

- SMPS topologies

Section 2.4:

- Sync buck operation principle
- Required gate driver timing
- Sequence of events during turn-on and turn-off
- Shoot-through effect
- Optimization of HS and LS die sizes

Chapter 3
Power MOSFET Performance

The first two chapters offered an introduction to power MOSFET fundamentals and described the expectation on MOSFET performance from the point of view of different applications. This chapter provides some basic hints how to make a good power MOSFET, especially one optimized for hard switching applications.

3.1 Breakdown Voltage and Internal Diode

Each design of a power MOSFET begins with the definition of the maximum allowable drain voltage for continuous operation ($V_{ds,max}$). The actual breakdown voltage of the device (BV) has to be higher than $V_{ds,max}$ by some margin. The margin value depends on the location of the breakdown voltage within the device. There are three typical locations of the BV:

– within each active cell at the corner of the gate,
– within each active cell away from the gate,
– at the perimeter of the active area of the device.

Breakdown location in the vicinity of the gate oxide is the most critical, as hot carriers generated during the breakdown avalanche may easily be injected into the oxide leading to so called "walk-out" of the breakdown characteristics, and finally to a physical breakdown of the oxide – i.e. to a blow up of the transistor.

Breakdown within the basic cell away from gate is the most robust, but requires dedicated drain engineering. Breakdown at the edge of the active area is the easiest to achieve, as the edge termination structures have inherently lower breakdown voltage capability than a flat PN junction. However, allowing the avalanche breakdown to concentrate within a narrow perimeter area will result in a restricted capability to absorb the avalanche energy.

Thus, the best engineering approach is to design a suitable edge termination structure which is similar to the breakdown capability of the MOSFET device at the corner of the gate, and design-in a lower breakdown voltage location within each active cell away from the gate. By doing so, an arrangement can be made

J. Korec, *Low Voltage Power MOSFETs*, SpringerBriefs in Applied Sciences and Technology 7, DOI 10.1007/978-1-4419-9320-5_3, © Jacek Korec 2011

which avoids hot carrier injection into the gate oxide. This way, sufficient avalanche robustness and good device reliability are secured.

Each MOSFET structure contains an internal diode created by the PN junction between body and drain regions. In the case of an N_{ch} transistor, body region has a P-type doping, and drain is doped N-type. So, applying a positive drain to source bias results in blocking of the internal diode, and the MOSFET conduction is controlled by the gate potential. When a negative bias is applied to the drain (3rd Quarter operation), the internal diode is conducting, and the voltage drop across the device can be reduced by parallel channel conduction induced by a positive gate bias.

A schematic diagram of a lateral MOSFET able to sustain a blocking voltage of interest is shown in Fig. 3.1. The blocking diode structure is created by P-body, P-well, N-LDD, and N-drain regions. The actual blocking voltage capability of the diode is defined by the doping and the length of the lightly doped drain extension (LDD). The electric field profile along the LDD region is shown in Fig. 3.2. This BV can be reduced on purpose increasing the doping underneath the drain contact – as discussed later.

The slope of the electric field distribution is proportional to the doping concentration, so it is highest within P-body, and N^+-drain regions. The lighter the doping of the LDD region the larger the area below the electric field distribution, hence, the larger the breakdown voltage of the diode. In a practical transistor structure, gate is overlapping the LDD region to assure channel coupling into the drain extension,

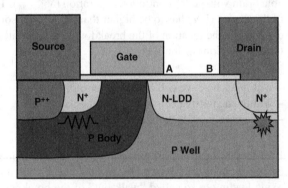

Fig. 3.1 Lateral MOSFET structure

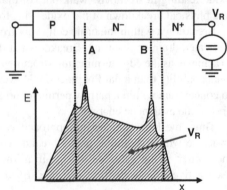

Fig. 3.2 Electric field within LDD region (Fig. 3.1)

and drain metal may overlap the LDD region creating a field plate action over a thin interposed oxide. The respective corners of the field plates (denoted by A and B) induce high electric field spikes in the field profile, and limit the achievable breakdown voltage value. For a given length of the LDD region there is an optimum doping concentration leading to similar heights of the voltage spikes at the A and B positions. If the doping is too high, breakdown voltage is limited by the electric field spike at the A position. In opposite, if the doping concentration is too low, B voltage spike is higher and limits blocking capability.

In general, in order to increase the BV of the lateral transistor, the LDD region has to be made longer, and the LDD doping has to be lighter. This of course results in an increase of the $R_{ds,on}$ of the device. Thus, the first goal of the MOSFET design is to minimize the on-resistance while keeping the target BV of the transistor.

Another region which needs a special attention of the designer is the perimeter of the active area of the device. If perimeter of a blocking PN junction is designed to simply terminate at the Silicon-Oxide interface, the device is not reliable. As soon as the electric field peak reaches the avalanche critical field value, hot carriers are injected into the oxide creating a trapped charge, and generating additional interface states. This leads to a walk-out of the blocking characteristics of the transistor. The simplest, and most popular, edge termination structure is a field plate as shown in Fig. 3.3.

The goal of the design of the edge termination is to stretch the electric field distribution along the Silicon-Oxide interface which results in a more uniform electric field profile and allows achieving higher breakdown voltage. There are many edge termination structures used for power MOSFETs, especially when target $V_{ds,max}$ is higher than 100 V, and a detailed discussion of this subject here is not possible. Actually, the simple field plate structure as depicted in Fig. 3.3 is good enough for low voltage devices, designed for operation at 30 V or less. The exact design can be easily optimized using numerical simulation techniques. However, as a rule of thumb, the oxide thickness underneath of the field plate (t_{ox}) should be close to a half of the depth of the PN junction. The length of extension of the filed plate beyond the edge of the PN junction (FP) should reach a value 1.5× to 3× larger than the junction depth.

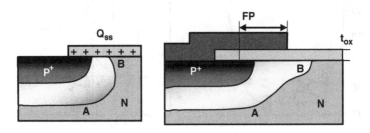

Design goal: $E_A \approx E_B$

Fig. 3.3 Electric field plate as a simple edge termination

As mentioned before, a good lateral MOSFET design makes sure that the device is going into breakdown at a specified location, preferably underneath of the drain contact as marked by a star mark in Fig. 3.1. The breakdown at this location has to be lower than the breakdown achievable at the edge termination, and at the LDD region. At the same time it is extremely important to avoid triggering of the parasitic NPN bipolar transistor built by the sequence of source/body/LDD regions. The minority carriers generated by avalanche (holes in N_{ch} MOSFET) at the breakdown location have to flow through the body region to the source contact. The minority current density multiplied by the sheet-Ro of the body region underneath of the source (see Fig. 3.1) creates a voltage drop biasing the source/body junction into forward direction. If this forward voltage drop exceeds about 0.6 V, source junction starts to inject electrons acting as emitter of the parasitic transistor. Triggering NPN transistor into action leads to a snap back of the blocking characteristics and to a destruction of the device. The remedy is to keep the length of the source region as short as possible, and to provide enough doping in the body region underneath of the source to limit the "base" resistor.

Finally, MOSFET designer has to care about the switching performance of the internal diode. This is especially important for applications using MOSFETs as a synchronous rectifier, i.e. like LS switch in the synchronous buck converter. Here, a free wheeling diode is substituted by a MOSFET, and a gate signal is applied to the switch to lower the conduction loss when the internal diode is supposed to conduct.

The internal diode conducts when the drain bias got reversed polarity. A stored charge is accumulated within the diode in proportion to the value of the flowing current. Some portion of the current is still conducted through the MOS channel in sub-threshold mode of operation. The lower the value of the V_{th} of the MOSFET, the less current is conducted by the diode, and so, the less stored charge has to be accumulated. Stored charge means that the concentration of mobile free carriers exceeds the doping level in some volume of the diode. The smaller volume of the lightly doped region designed to support the blocking voltage, the smaller the associated stored charge. Concentration profile of the stored charge during a commutation of the diode is shown in Fig. 3.4.

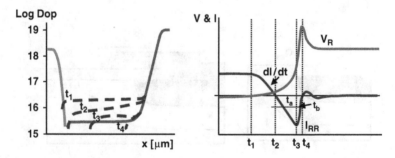

Fig. 3.4 Illustration of forced commutation of a PN diode

Figure 3.4 shows free carrier concentration profiles at times t_1 through t_4 as marked in the plot of switching I-V waveforms. Time t_1 denotes the situation of diode conduction prior commutation. Forced commutation means that the voltage drop across the diode is reversed by the external circuit, and the diode has to adapt to it by changing from conduction to blocking state. The slope of the current change in response to the changed bias (dI/dt) is defined by the value of the applied new bias and the value of the parasitic inductance of the commutating circuit. At time t_2 the concentration level of the free carriers in the vicinity of the PN junction drops down to the level of the lightly doped region, the PN junction emerges and is getting ready to develop a depletion region needed to support a blocking voltage. At this point of time the direction flow of the current changes the diode begins reverse recovery. At time t_3 the depletion region is developed fully to the dimension needed to sustain the applied blocking voltage. Further current flow has to be provided by the rest of the stored charge allocated beyond the depletion region. If there is not enough stored charge left, the current flow decays abruptly leading to a high voltage spike in the circuit. Such diode is called snappy. The level of snappiness is measured by the ratio of time intervals t_b/t_a (Fig. 3.3). Some time interval t_b is always there, as the diode capacitance (output capacitance of the MOSFET) has to be charged to accommodate the new voltage drop. Soft switching diodes have a ratio t_b/t_a larger than one. Soft diodes have to be used in applications where large voltage spikes are not acceptable.

The reverse current peak value I_{rr} depends on the slope dI/dt and the amount of charge stored in the diode during the conduction stage. The amount of area enclosed by the current recovery waveform determines the stored charge Q_{rr} published in MOSFET data sheets. In push-pull configurations which are very popular in DC/DC converters, the Irr current caused by the internal body diode of the LS-switch is adding to the current flowing through the HS-switch which increases HS switching loss. This discussion will be continued in the chapters to follow.

3.2 R_{ds,on} Components and V_{th} Impact

$R_{ds,on}$ components in a Trench-MOSFET are sketched in Fig. 3.5. Typically, in a 20–30 V designed MOSFET, metal and contact resistive contributions are small. Thus, channel, epitaxial layer, and substrate resistances are similar, contributing about 30% to total $R_{ds,on}$ each. In a case of a different power MOSFET structure, like for example in a case of a lateral MOSFET from Fig. 3.1, the on-resistance contribution from the epitaxial layer is substituted by the contribution from LDD region, but the general picture stays the same.

When comparing MOSFETs designed for different BV's, the contribution of the lightly doped region which is supporting the blocking voltage increases with increasing targeted $V_{ds,max}$, as shown in Fig. 3.6. Channel exhibits only a small increase with target BV, and other contributions don't scale.

Fig. 3.5 R$_{ds,on}$ components
in a Trench-MOSFET

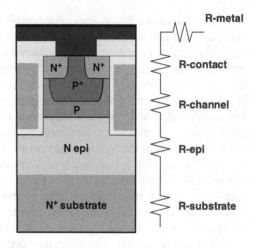

Fig. 3.6 Specific R$_{ds,on}$ vs.
target breakdown voltage

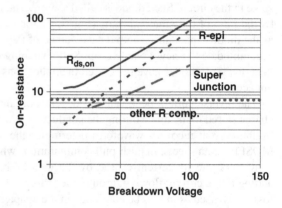

Two theoretical limits are shown in Fig. 3.6. The increase of the resistance of the
lightly doped region supporting the blocking voltage – epitaxial layer in case of a
Trench-MOSFET (R-epi), follows the design rule for depletion region length and
doping in a planar PN junction with uniform doping. The second limit relates to a
technology called super-junction, where depletion region develops within a volume
composed of regions with alternating doping types (N and P). The doping con-
centrations in these alternating regions are close, so that the charge of the ionized
donors and acceptors cancel each other within depleted volume. The super-junction
technology became state-of-art for MOSFETs designed for medium voltages (100–
1,000 V), and has been even applied in products designed for 60 V. The advantage
diminishes with the designed BV, and the added complexity of the processing
doesn't pay back for products designed for 30 V and less.

Further development of low voltage MOSFETs in terms of R$_{ds,on}$ minimization
is focused on increasing channel density (smaller basic cell pitch results in smaller
specific R$_{ds,on}$), and reduction of substrate contribution by use of highly doped, thin

Fig. 3.7 $R_{ds,on}$ vs. V_{gs} for
two technologies with
different V_{th} and g_m

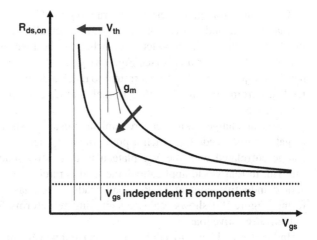

substrates. Otherwise, the design of MOSFETs for voltages below 20 V is taking
advantage of shorter channel length, thinner oxide (high g_m), and low V_{th}.

As shown schematically in Fig. 3.7, lower V_{th} reduces $R_{ds,on}$ at low V_{gs}, and
higher transconductance (g_m) leads to a faster drop of $R_{ds,on}$ for V_{gs} higher than
V_{th}.

It has to be noticed that low $R_{ds,on}$ is all what matters for load switch applications.
On the other hand, switching performance is at least as important for PWM applications. However, specific $R_{ds,on}$ is still important as the prime factor determining the
die cost for a specified target on-resistance of the product. To take into consideration
both aspects, two Figure of Merit (FOM) factors are used to benchmark MOSFET
products, as described in the next chapter.

3.3 MOSFET Switching

Switching performance of a MOSFET is dictated by the value of built-in capacitances. Shortly speaking, the design of MOSFETs with minimum internal capacitances is a must for hard switching applications. There are three capacitances
incorporated in a MOSFET structure, as shown in Fig. 3.8.

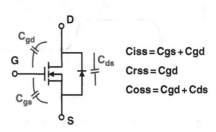

Fig. 3.8 MOSFET
capacitances

C_{gd} results from gate overlapping drain region, C_{gs} respectively is caused by the overlap of gate and source region, and C_{ds} is actually the capacitance of the built-in PN-junction. At the product level, what can be measured are the capacitances between the three terminals (see Fig. 3.8). C_{iss} defines the amount of charge which has to be injected into the gate terminal to reach a desired V_{gs} level. C_{rss} determines feedback from drain to the gate terminal, and C_{oss} is simply the MOSFET output capacitance.

Two gate charge parameters are published in a data sheet. Q_g defines the amount of gate charge needed to reach a pre-defined V_{gs} level, and Q_{gd} is the amount of charge correlated with the V_{gs} plateau in a resistive load switching curve – see Fig. 1.6. In real world applications the load is rarely pure resistive. The capacitive and/or inductive content of the load leads to a phase shift of the I-V waveforms. Figure 3.9 shows realistic switching waveforms for the case of a mixed inductive/resistive load.

Inductive load leads to a delay of the current waveform versus the voltage waveform as compared to the resistive case. This is the cause of a smaller I-V overlap during turn-on than during turn-off, and leads to a respective modification of the switching power losses. Figure 3.9 shows switching waveforms for two similar MOSFETs. The difference is that the second MOSFET has double C_{gd} and C_{gs} values resulting in slightly slower switching.

The first MOSFET being a faster device induces a higher dV/dt at the rise V_{ds} edge. This in turn ends up in pushing the MOSFET into avalanche which clamps V_{ds} at the BV of the switch. The second MOSFET being a slower switching device induces a smaller V_{ds} spike, but shows slightly larger area of the overlap of I-V waveforms which means larger switching loss.

Avalanche event in a MOSFET under breakdown condition has to be carefully studied by the designer as it can lead to a destructive triggering of the parasitic

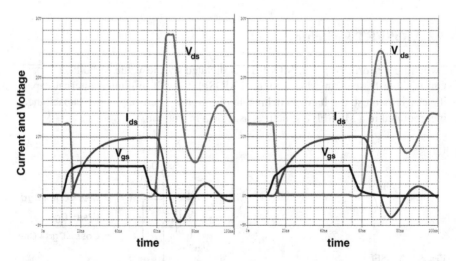

Fig. 3.9 Inductive load switching waveforms for two MOSFETs

Fig. 3.10 Avalanche
locations in a
trench-MOSFET

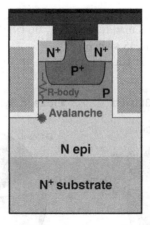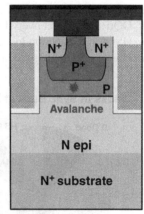

bipolar transistor. This topic has been shortly discussed in Section 3.1 (Figs. 3.1 and 3.2) and is once again illustrated by Fig. 3.10 for the case of a Trench-MOSFET.

Avalanche breakdown occurring at the bottom corner of the trench presents a reliability issue as hot carriers will be injected into the gate oxide. A better design is to place the avalanche location just underneath of the body contact in the middle between the trenches (Fig. 3.10). This is possible by a careful doping engineering, but is increasingly difficult with cell pitch getting smaller. Actually, it can not be done anymore if a high trench density is applied and the cell pitch goes below 1 μm.

3.4 Power Loss Components

The total power loss in a switching application consists of conductive and switching power loss. Additionally, as mentioned before, the specific $R_{ds,on}$ of the MOSFET determines the cost of the die which is proportional to its area. For benchmarking purposes two FOM's (figure of Merit) have been introduced:

$$FOM1 = R_{ds,on} {}^*Q_g, \text{ and } FOM2 = R_{ds,on} {}^*Q_{gd}.$$

The first one reflects conductive loss and the respective driver loss, and the second one reflects conductive and switching loss. In a case of a synchronous buck converter, FOM1 is suitable to benchmark the usually large area LS device, and FOM2 is more appropriate for the HS switch.

An example of a power loss breakdown for a synchronous buck operating at two different switching frequencies is presented in Fig. 3.11. The major power loss contributions at 300 kHz are:

- LSon: conduction loss in LS switch
- LSQoss: loss due to charging of the LS output capacitance

- Dcon: conduction loss of the body diode
- DQrr: diode recovery loss manifesting as I_{rr} flowing through HS switch
- HSon: conduction loss in HS switch
- HSsw: switching loss in HS device

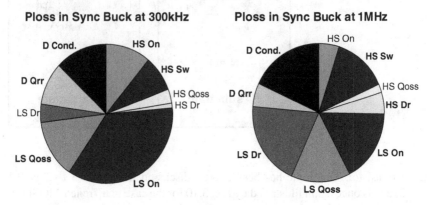

Fig. 3.11 Power loss breakdown in a synchronous buck converter

When moving to a higher switching frequency of 1 MHz, the relative weight of the individual contributions shifts to switching losses which become more important. Also losses associated with gate drivers (LSDr and HSDr) get significant.

From the designer point of view, following MOSFET parameters have to be optimized:

- small $R_{ds,on}$ ->   reducing conduction losses
- small Q_g ->   reducing driver losses
- small Q_{gd} ->   reducing switching losses
- small Q_{rr} ->   reducing HS losses due to I_{rr} in LS body diode

Excessive diode conduction losses can be avoided if the delay time between HS and LS signals is kept minimal (see Fig. 2.18) or if a Schottky diode is integrated with the MOSFET in parallel to the internal body diode to reduce the forward voltage drop during diode conduction.

What can be done in the design of the sync buck converter to minimize the power loss, i.e. to increase power conversion efficiency, is discussed in some more detail in the last chapter.

Chapter Summary

Following topics have been handled in this chapter:
Section 3.1:

> - MOSFET design for a target BV
> - Edge termination

- Avalanche breakdown
- Parasitic bipolar transistor

Section 3.2:
- $R_{ds,on}$ components
- V_{th} and g_m impact on $R_{ds,on}$

Section 3.3:
- MOSFET switching performance
- Impact of internal capacitances

Section 3.4:
- Power loss components
- Figures of Merit
- Impact of switching frequency

Chapter 4
MOSFET Generations

The introduction of a new MOSFET generation on the market usually takes place in a situation when the previous generation has been there for many years and further incremental improvement of the technology doesn't provide satisfactory improvement of the MOSFET performance. Figure 4.1 illustrates how the introduction of every new MOSFET generation in the past significantly improved the FOM. In marketing papers the transition steps are called technology breakthrough.

The reasons behind performance improvement and the remaining issues involved in the consecutive MOSFET generations are discussed below.

Fig. 4.1 FOM improvement by introduction of new MOSFET generations

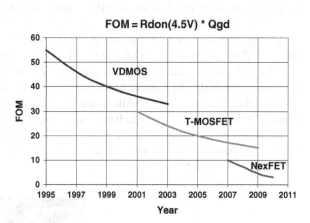

4.1 LDMOS and VDMOS

Lateral Double Diffused MOSFET (LDMOS) was the first power MOSFET structure which is still used to build output power stages in power management IC's. As discussed in Section 2.1, a lateral structure of power MOSFET is not adequate to design switches for high currents, where a vertical transistor should be used instead. The technical issues inherent to a lateral MOSFET design, as shown in Fig. 3.1, have been alleviated by technical improvements shown in Fig. 4.2.

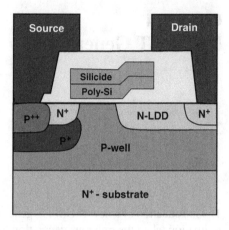

The high electric field spike occurring at the corner of the gate can be reduced by
inserting a thicker oxide made by LOCOS oxidation, and the resistance of the base
region of the parasitic bipolar transistor can be reduced by an additional implantation to increase the dopant concentration underneath the source region. One of the
remaining design issues is the fact that the lateral transistor has to be created using
the existing doping and mask steps in the manufacturing flow of the IC. As a result
the on-resistance and the internal capacitances of the LDMOS are not optimal.

To improve the current capability of the power MOSFET, a Vertical Double
Diffused MOSFET structure (VDMOS) has been introduced in the early 1980s.

A basic VDMOS structure is shown in Fig. 4.3. A large area top electrode is
the source terminal. The current flows from the source region N$^+$ to the channel
underneath of the lateral gate, and then into the N epitaxial layer where it is diverted
to the N$^+$ substrate and finally reaches the drain electrode at the back of the die.

The main design issue inherent to the VDMOS structure is current pinching
between the two P-body regions as illustrated by Fig. 4.4.

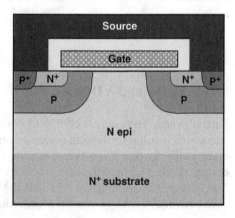

Fig. 4.3 Structure of a
VDMOS transistor

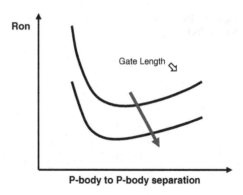

Fig. 4.4 JFET effect in a VDMOS transistor

Current pinching called JFET effect takes place if the distance between the two P-body regions is too small. This effect limits the ability to reduce the cell pitch in the design of a VDMOS structure. As mentioned before, reducing pitch is the easies way to reduce the specific on-resistance of the transistor. The other way is to make channel shorter, but this is limited by the danger of punch-through. Punch-through occurs when the depletion induced by drain voltage penetrates body region and reaches the source N^+ region. Another issue is created by the overlap of the gate on drain region between the P-body wells. The gate overlap leads to an increased C_{gd} capacitance which can not be minimized due to the presence of the JFET effect. Thus, there is only a restricted chance of improving FOM of the VDMOS design.

4.2 Trench MOSFET

Trench MOSFET has been introduced in the 90s to avoid the restriction in improvement of the FOM in a VDMOS structure. Here, the density of the channel per unit area can be increased dramatically. The result is a proportional decrease of the on-resistance accompanied by an increase of the capacitance due to the large area of the trench wall. As illustrated by Figs. 3.5 and 3.6, the reduction of the on-resistance encounters a limit imposed by the non-scalable resistance components like contact or substrate resistance. On the other hand, Q_g and Q_{gd} keep increasing linearly with the reduction of the pitch of the basic cell. This means that from the point of view of switching applications there is an optimum trench cell density leading to an optimum FOM. From the point of view of load switching (minimum $R_{ds,on}$) the trench density is limited by the patterning ability of the used technology.

The other basic design issue concerning the location of the avalanche breakdown in a Trench-MOSFET was discussed in conjunction with Fig. 3.10. Rounding of the bottom trench corner and a thicker gate oxide at the bottom of the trench has been used to improve the long term reliability of Trench transistors. The later has been also used to reduce the C_{gd} defined by the gate/drain overlap along the bottom surface of the trench. Another way to improve FOM of modern Trench-FETs

Fig. 4.5 Trench-MOSFET
with a shielding gate structure

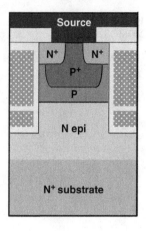

is to split the Polysilicon in the trench into two gates: the top one acting as the
control gate, and the bottom segment being connected to the source shields the top
gate from the drain potential. This approach called shielding gate is illustrated by
Fig. 4.5.

The shielding gate is very effective in reducing C_{gd}, but increases the cost of
processing significantly. An alternative approach is just to use narrow trenches with
minimum area of the trench bottom. Also, because of the bad control of the length
of overlap of the gate extension beyond body region, C_{gd} is still relatively high no
matter whether shielding gate or narrow trench approach is used. Trying to use short
channel approach, the dopant concentration in the body region has to increase what
leads to a high threshold voltage of the transistor. In other words there are design
limitations inherent to the Trench MOSFET structure which can't be removed easily. Thus, Trench-FET development is in a matured stage of technology optimization
and further development can be expected to bring an incremental improvement
only.

4.3 NexFET

NexFET is a new approach to power MOSFET design and has its roots in an
LDMOS based transistor optimized for RF applications. The basic structure is
shown in Fig. 4.6. The issues related to a design of a FET with a planar gate like in
Fig. 3.1 are resolved by some structure modifications.

As shown in Fig. 4.6, the source metal creates a large area top electrode, and the
current flow is diverted from the lateral channel into the substrate to reach the drain
electrode at the back of the die. Vertical current flow makes this device suitable for
carrying high current density. Additionally, the top source metal overlaps Polycide
gate stack and comes close to the substrate surface on the LDD side of the cell.
This concept has two functions: first the gate is shielded from the drain potential

Fig. 4.6 Basic structure of a
NexFET transistor

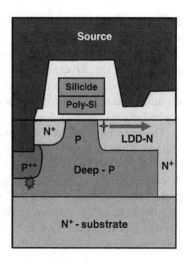

appearing in the vertical current plug N^+, and second, the portion of the source metal overlying LDD creates a field plate action stretching the electric field distribution away from the gate and reducing the critical electric field spike at the gate corner. Also, even a small value of the drain bias depletes the LDD region under the filed plate what results in a rapid decrease of the C_{gd} capacitance.

The high dose implantation through the bottom of the shallow source trench makes a good ohmic contact to the body region, but also lowers the breakdown voltage at this location. The location of the breakdown just underneath of the source contact makes the device quite rugged in terms of absorption of a large amount of avalanche energy.

The doping of the LDD region follows the design guideline for charge balanced regions. Under depletion condition created by a drain blocking voltage, the LDD charge is balanced by the depletion reaching into the deep-P region, and by the field induced by the source metal field plate region. This approach allows an increase of the doping concentration in the LDD region by almost one order of magnitude as compared to a non-optimized lateral MOSFET design from Fig. 3.1. Charge balance design similar to the super-junction approach allows to reduce the pitch of the cell, and to minimize the on-resistance of the transistor.

The process flow is specifically optimized for the manufacturing of a discrete product, so the gate portions overlapping the source and LDD regions are minimized which in turn makes C_{gs} and C_{gd} values very small. Adjusting body implantation conditions, the doping profile in the body region along the interface with the gate can be made flat which allows a reduction of the channel length while keeping low V_{th} value.

NexFET approach allows to create products with competitive specific $R_{ds,on}$ values while showing significant advantage in Q_g and Q_{gd} values. Respectively, achievable FOM values are better by a factor of $2\times$ to $3\times$ as compared to the state-of-art Trench-MOSFETs.

Better FOM values illustrate the ability of NexFETs to switch faster with lower power loss at the driver side. Data published in the literature show that the DC/DC converters using NexFET transistors as switches achieve highest efficiency levels and/or allow increasing the switching frequency within the same thermal limit applied to the converter design.

The advantage of NexFET switches increases with decreasing target $V_{ds,max}$ voltage. This allows closing an existing gap in voltage range of power MOSFET products. LDMOS transistors using VLSI technology can be expanded up to 7–9 V applications, but then their performance is getting bad. Trench-MOSFET performance improves when scaling the design down to about 20 V, but further improvement below 20 V is hardly achievable. In contrast, the design of NexFET devices has its sweet spot voltage range between 25 and 10 V, with improved performance when lowering BV.

Chapter Summary

Following topics have been handled in this chapter:

Section 4.1:

- LDMOS and VDMOS features

Section 4.2:

- Advantage and issues of a trench MOSFET structure

Section 4.3:

- What is different in a NexFET

Chapter 5
Sync Buck Optimization

This is the last chapter of the script discussing the ways how to take advantage of the improved performance of state-of-art MOSFETs as presented above. From the point of view of switching applications the most important improvement has been achieved by minimizing the switching power loss. Less switching loss allows an increase of the switching frequency of an SMPS unit, which in turn can be used to improve performance of the system as discussed in Section 2.2 and illustrated in Fig. 2.11.

Another way of taking advantage of the lower power loss is to keep the same switching frequency of the converter and enjoy a higher efficiency of power conversion as discussed in Section 2.4 and shown in Fig. 2.20.

Further in depth analysis how to improve the performance of a synchronous buck converter will be done below. Special attention will be given to the matching of the gate driver to the chosen MOSFET chip set, assuring dV/dt immunity and keeping the switch node ringing under control.

5.1 Gate Driver

The principle of operation of a synchronous buck converter has been explained in Section 2.4 and illustrated with Figs. 2.17 and 2.18. The basic role of the gate driver is to use the clock PWM signal supplied by a controller IC to generate gate signal sequence for HS and LS switches, and to supply enough current to MOSFET gates to switch them fast. A closer look at the timing sequence of the gate signals is shown in Fig. 5.1.

The delay introduced by the gate driver between the turn-off of the LS switch and the turn-on of the HS switch has to be long enough to avoid cross current. LS MOSFET being usually a larger device has a longer internal delay time than the small HS die. This means that the actual gate waveforms may overlap even in a presence of a delay time between the gate driver signals. However, as long as the LS-V_{gs} drops below V_{th} before the HS turns-on, the cross current is avoided, and a dead time in the conduction of the two switches is introduced. The free wheeling current is conducted by the body diode of the LS transistor during this time. A diode

J. Korec, *Low Voltage Power MOSFETs*, SpringerBriefs in Applied Sciences and Technology 7, DOI 10.1007/978-1-4419-9320-5_5, © Jacek Korec 2011

Fig. 5.1 Gate signal timing sequence at the rise edge of the switch node voltage

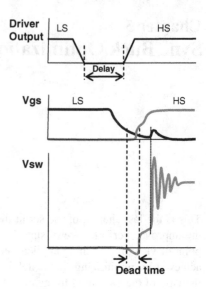

conduction loss is generated due to the significant forward voltage drop of the PN diode. Thus, an optimum matching of the gate driver signals to the implemented MOSFET chip set will be achieved when the dead time is close to zero.

If the driver IC is designed for a large family of MOSFETs, an exact matching of the delay time can not be achieved. Following approaches have been adopted:

– fixed delay time which can be adjusted by an external resistor or capacitor,
– adaptive delay time adjusted by monitoring the status of the LS switch,
– predictive approach where the delay time is adjusted based on the dead time in the previous cycle.

Because the adjustment range of a fixed delay time and its accuracy are not good enough, the implementation of this technique usually results in relatively long dead times with negative impact on the converter efficiency. Adaptive delay time is a popular approach, where the voltage drop across the LS switch or the actual gate bias at the transistor are monitored, and the HS signal is generated as soon as the monitored voltage drops below a pre-defined value. This is a good approach, and the only issue is the actual propagation time of the feedback signal and the reaction time of the gate driver IC. As a result, dead times shorter than 10 ns are difficult to achieve. Predictive approach is based on monitoring the time period when the voltage drop across the LS switch is negative (diode conduction) and respective cut back of the delay time for the next switching cycle. This sounds like a smart approach, but it can work properly only if the switching cycle period is short as compared with speed of change of the load conditions. The predictive approach has not been accepted in computer applications like power supply for central processor unit (CPU) where the load condition is changing frequently and abruptly.

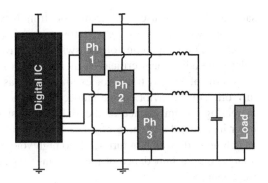

Fig. 5.2 Multi-phase sync buck converter topology

As discussed above there is no chance for tight delay time optimization in implementation based on discrete components, especially if the components are randomly chosen "from the shelf". An exact matching of the driver timing to the specific MOSFET chip set can be achieved if all components are known and are integrated into a functional block implementing the power switching action between the control IC and the output filter (Fig. 5.2).

The functional blocks called power stages offer an easy approach to implementation of multi-phase converters where the phase shift of on-cycles of the individual phases allows an increase of the effective switching frequency of the converter and a respective optimum design of the output filter including coupled inductors and ceramic output capacitors. Power stages also allow an easy implementation of protection and monitoring functions which are essential in realization of digitally controlled systems. The author is convinced that this is the approach which will dominate the sync buck architectures in the future.

The integration of gate driver with MOSFET switches into one module has the big advantage that dynamic behavior of the selected components is known and the gate signal timing can be matched precisely. The preferred length of the dead time is about 5 ns which is long enough to provide a head room needed to avoid a cross current, and is short enough in the sense that the internal body diode doesn't have time to build up any significant stored charge. As described in Section 2.4, small stored charge means small reverse recovery peak, smaller switching loss of the HS transistor and smaller ringing of the switched node. However, the length of the dead time changes for a given delay time when the operating point shifts from light to heavy load. This means that an optimized gate driver should monitor the output current, include a lookup table and adjust the delay time accordingly.

Another important issue is related to the current capability of the output stage of the driver. The gate drivers introduced in the last decade have been designed to work well with Trench MOSFETs which have large input capacitance. For example, in order to switch a transistor with C_{iss} in the range of 2–3 nF, and achieve switching times below 10 ns, the current capability of the gate driver has to be 3 A or higher. Modern gate drivers offer current capability of 5 A for both charge and sink currents to drive fast heavy Trench MOSFETs designed for switching 20 A or more. If the same gate driver is used with transistors from NexFET family, the result will be fast switching with sharp voltage waveforms, and respectively prolonged dead time.

Sharp switching edges of the switching waveforms make the precision of the delay time adjustment critical. A more tolerant system can be achieved if the gate waveform of the LS transistor exhibits a tail as the bias drops below V_{th} during turn-off. This behavior can be induced by reducing the sink current, this means by increasing the impedance of the sink output stage of the LS gate driver. On the other hand, hard switching of a fast HS MOSFET leads to very high dV/dt at the switching node which in turn leads to high voltage ringing as discussed later. Here, a larger output impedance of the charging stage of the HS gate driver may be the right approach.

A small shoulder in the V_{sw} waveform appearing just after switch node voltage becomes positive is related to the diode reverse recovery current flowing through the parasitic inductance attached to the drain of the LS switch. After the diode has removed the stored charge, the growing depletion region can accommodate a rapidly increasing drain voltage (Fig. 3.4) and a high dV/dt is observed. This in turn can lead to a shoot-through event as described in Section 2.4.

Similar consideration can be made for the falling edge of the switch node voltage. However, as long as the LS switch is larger or equal in the die size to the HS transistor, there is little concern about simultaneous conduction of HS and LS switches, and the delay time of gate signals can be pretty short.

In summary, talking about gate driver optimization makes sense only if we know all the switching components and have under control the parasitic inductances introduced by PCB layout and assembly. This is feasible for integrated power stage modules, but can hardly be achieved in discrete implementation.

5.2 dV/dt Immunity

There are two events which may affect dV/dt immunity of a MOSFET working in a synchronous buck converter. Both apply to the LS switch:

– triggering of the parasitic bipolar transistor during the commutation of the body diode,
– shoot-through current induced by a bounce of the LS gate voltage above V_{th}.

The mechanism of triggering of the parasitic bipolar transistor has been explained in Section 3.1 as referred to Fig. 3.1. From application point of view there is not much what can be done to avoid the danger of bipolar action. As the internal body diode of the LS switch conducts during the dead time it stores some excess in concentration of mobile carriers. This stored charge has to sink to the source contact as soon as the voltage across the LS transistor is reversed. The induced current density flowing through the body region multiplied by the resistance of the body region underneath the source region creates a voltage drop biasing the source to body diode in forward direction. To avoid bipolar transistor trigger action, forward bias at the tip of the source region has to stay well below 0.6 V. Robust MOSFET design includes:

– low Q_{rr} body diode,
– short distance between source contact and gate (Fig. 3.1),
– high doping concentration in body region underneath of source.

Of course, dead time close to zero avoids diode conduction – what means it prevents accumulation of stored charge in the diode and the resulting danger of bipolar action.

The generation of a shoot-through current has been described in Section 2.4 and illustrated by Fig. 2.19. The height of the voltage bounce at the LS gate terminal can be modeled with an analytical formula which is illustrated by Fig. 5.3.

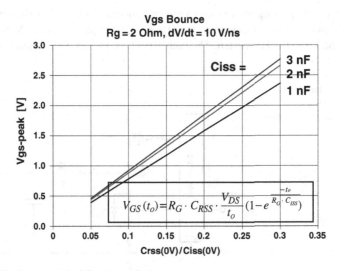

Fig. 5.3 V_{gs} bounce at the LS gate terminal

The model calculates V_{gs} peak after 1 ns of dV/dt action and takes into consideration C_{rss} and C_{iss} capacitance values at zero drain voltage, and a gate resistance of 2 Ω including the output resistance of the sink stage of the gate driver. The plot can be read in the way that the C_{rss}/C_{iss} ratio should be below 0.15 if a shoot-through effect should be avoided for a transistor V_{th} of 1.5 V and a gate path resistance of 2 Ω.

It has to be noticed that an increase of the output sink resistance of the gate driver which was proposed to close the dead time gap (Fig. 5.1) and eliminate diode conduction and Q_{rr} related problems is allowed only in a case where the LS device in use is shoot-through immune. Otherwise following rules have to be observed:

– use MOSFETs with small C_{rss}/C_{iss} ratio,
– be sure that MOSFET V_{th} is sufficiently large,

– limit dV/dt by one of the techniques discussed in the next chapter,
– add some external C_{gs} to the LS switch.

In practical applications some small degree of shoot-through is tolerable, but it
lowers the converter efficiency at light load condition (Fig. 2.20). Also, during the
shoot-through event a small current flows through the LS transistor which already
blocks a significant V_{ds} voltage. This may lead to reliability concerns even if the
operating point is well within the allowed SOA regime (Safe Operating Area). For
these reasons it is wise to avoid any shoot-through current if possible.

5.3 Mastering of Ringing

Switch node voltage ringing similar to the one shown in Fig. 3.9 is the result of a
resonant response of the components in the main current path to a current commutation from the LS to the HS switch. The components of interest are shown in Fig. 5.4.

Fig. 5.4 Sync buck converter
including parasitic
components due to assembly

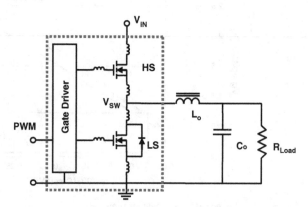

The power stage module discussed before is marked in Fig. 5.4 by a dotted
line. The main current path during switching is between the supply voltage V_{in}
and ground. Respectively, the resonant tank is built up by the output capacitances
of the HS and LS MOSFETS in conjunction with the parasitic inductance of the
PCB layout and power stage assembly. It is obvious that an integrated power stage
module will have much smaller parasitic inductance than a circuit built with discrete
components where the inductance of the routing is adding to the sum of inductances
of individual packages. Thus, the use of integrated power modules is advisable also
from the point of view of mastering of the ringing issue.

The ringing event occurs in following steps:

– HS transistor turns-on supplying current to absorb I_{rr} of the LS diode and charge
  the output capacitance of the LS switch,

- initial dI/dt stores energy in the parasitic inductances attached to source and drain terminals,
- proportionally to the injected Q_{oss} into the LS switch, the LS-V_{ds} increases, and the HS switch starts to supply current to the output filter,
- HS current increases rapidly leading to large dV/dt at the switched node, and results in further energy storage in the parasitic inductances attached to the HS terminals,
- large dV/dt may induce shoot-through event in the LS switch, where the corresponding shoot-through current has to be also supplied by the HS switch,
- energy stored in the resonant tank between V_{in} and ground leads to voltage oscillation which is damped by the resistance of the current path.

So, the ringing event can be observed in two main steps. The first is the energy storage in the resonant tank with the related first voltage spike, and the second is energy oscillation with the corresponding voltage ringing and damping proportional to energy losses. There are two corresponding ways to deal with the ringing issue. The first is to limit the amount of energy stored during commutation (first voltage spike), and the second is to damp the following energy oscillation (further voltage ringing).

Damping of the ringing is trivial, and is usually done by different snubber techniques. The most straightforward one is to place an RC snubber between the switched node and ground. The related issue is coming from the fact that damping is just energy dissipation: the more damping the less efficient converter is the result. Also, an unacceptably large snubber would have to be used to lower the height of the first spike, so a different approach has to be used. In opposite, the smaller first spike is induced, the less energy is present in the resonant tank, and proportionally smaller snubber is sufficient.

The main tool used to limit the height of the first voltage spike is to slow down the switching speed of the HS MOSFET. Restricting the current amount delivered by the HS switch at the start of the commutation limits the dV/dt which is built across the C_{oss} of the LS switch. The rest is straightforward: the smaller dV/dt, the smaller is the height of the first voltage spike, and the less damping is required for the following voltage ringing. There are different options how to slow down the switching speed of the HS device during turn-on:

- attach an external R_g to the gate terminal of the HS switch. This approach leads to a large efficiency loss as the switching speed is reduced for both turn-on and turn-off of the HS transistor and the resulting switching power loss is too large.
- introduce a resistor (R_{boot}) between the bootstrap capacitor and the supply rail for the HS gate power output of the gate driver. This approach is better than the previous one as it affects turn-on speed only, but it creates a dip in the voltage supply to the HS gate driver which may end up in instable operation.
- increase the output resistance of the charging stage of the HS gate power output allows adjustment of the speed of the turn-on leaving the fast turn-off untouched.

The third option is clearly the best one, but it requires a dedicated design of the gate driver which is usually acceptable for integrated modules only.

As the reverse recovery current of the body diode is comparable to the amount of displacement current needed to charge C_{oss} of the LS transistor, a large reduction of the first voltage spike is achieved using MOSFETs with less Q_{rr}. One popular approach is to use MOSFETs with integrated Schottky diode which clamps the internal body diode during conduction and limits the related stored charge. This is a good solution for discrete assembly and low switching frequency. As the switching frequency approaches 1 MHz, the use of integrated power stage modules with less parasitic components and an optimized gate driver will prevail. In this case, cutting the dead time down to a few nanoseconds will limit the impact of the diode conduction and recovery anyhow.

Chapter Summary

Following topics have been handled in this chapter:

Section 5.1:

- Selection and matching of the gate driver
- Multi-phase sync buck topology

Section 5.2:

- Possible issues with fast dV/dt transients
- How to assure good dV/dt immunity

Section 5.3:

- Parasitic components and switch node voltage ringing
- Sequence of events leading to switch node voltage ringing

Chapter 6
High Frequency Switching

The meaning of the term high frequency switching as related to a synchronous buck converter applications depends on the power range, the step in the voltage conversion and the switching performance of MOSFETs used as power switches. In a power range of 1 W, and low input voltages, a switching frequency up to 10 MHz can be reasonably achieved. In contrast, switching 20 A from V_{in} of 12 V down to 1.2 V at a frequency of 2 MHz is a big hurdle if an efficiency close to 90% is expected. Switching power loss increases with output current and input voltage level, and is proportional to switching frequency. Respectively, novel, fast switching MOSFETs have to be used. On the other hand, 2 MHz frequency means a period of 500 ns only. If a duty cycle of 10% is required, the whole on-cycle has to be accomplished within 50 ns. This requires very effective and fast gate voltage control.

6.1 General Issues

In a gate charging circuit as shown in Fig. 6.1 attention has to be made to keep gate and source inductances at minimum, and a low R_g value allows fast MOSFET switching.

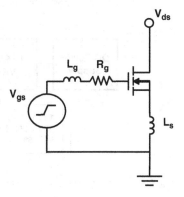

Fig. 6.1 Gate charge circuit
with parasitic components

J. Korec, *Low Voltage Power MOSFETs*, SpringerBriefs in Applied Sciences
and Technology 7, DOI 10.1007/978-1-4419-9320-5_6, © Jacek Korec 2011

If the parasitic components are not negligible, following effects will be observed:

– gate inductance slows down charging of the MOSFET gate and speeds up the discharge process,
– R_g limits gate current and slows down gate control,
– source inductance responds with a voltage bounce to the dI/dt during turn-on and slows down the gate charging speed.

In end effect, a hard switching can not be accomplished with a low V_{gs} supplied by the gate driver. Also, the inductance attached to the MOSFET source has to be minimized.

The new MOSFET generations on the market combined with the use of integrated power stage modules with optimized gate drivers make switching frequency of 5 MHz feasible, especially if a two stage approach is adapted (Fig. 6.2).

The first VRM stage performs a large voltage step-down to an intermediate voltage value which is loosely controlled. The second stage operates from a low input voltage, so it can easier achieve high switching frequency with acceptable power loss. Also, the switches used in the second VRM stage can be designed for low BV voltage, which helps in achieving good MOSFET performance.

In the following, the impact of using low voltage switches on high frequency performance of a synchronous buck converter will be discussed. Two scenarios will be considered:

Scenario I:

–   Cost of MOSFETs is reduced by using lower R_{sp} to cut down the die size.

Scenario II:

–   Total Si area is kept constant and partitioned as follows:

$$LS\_area / HS\_area = SQRT((1 - DC)/DC)$$

–   Lower power loss is used to increase switching frequency.
–   System cost is reduced by reduction of the size of the output filter.

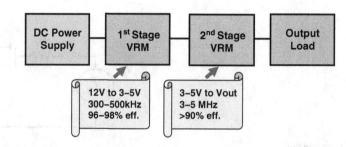

Fig. 6.2 Two stage sync buck concept

For the sake of this discussion power loss in the sync buck converter is calculated using a simplified set of equations presented in Appendix A. In both scenarios the reference data are obtained for the case of 1st and 2nd stage MOSFETs designed for BV of 30 V. The advantage of using a low intermediate voltage between the 1st and the 2nd stage is shown by redesigning the MOSFETs used in the 2nd stage for lower breakdown voltage values. All data presented for Scenarios I and II have been collected for one chosen technology of advanced power MOSFETs.

6.2 Impact of Low V_{IN}: Scenario I

Low value of the intermediate voltage allows a redesign of MOSFETs used for the second VRM stage. Low input voltage for the 2nd stage means, that the switches can be designed for low breakdown voltage (BV) which significantly improves their specific on-resistance ($R_{sp} = R_{ds,on}$ * Area). Using 30 V MOSFET design as a reference the reduction of R_{sp} for lower voltage devices is shown in Fig. 6.3.

A redesign of a MOSFET for a lower BV results in a smaller pitch of the active cell increasing the density of the channel per unit area of the Silicon die. This is one of the main reasons for the reduced R_{sp} value. Additionally the doping of drift region (or L_{DD}) can be increased, and the thickness of the epitaxial layer can be reduced further lowering R_{sp}.

At the same time, the higher density of cells leads to an increase of the internal capacitances manifesting themselves in higher Q_{xx} values (Figs. 6.4 and 6.5).

In the first Scenario the advantage of designing the 2nd stage MOSFETs for lower breakdown voltage is used to lower the cost of the power switches by reduction of the die size area while keeping the same $R_{ds,on}$ of the HS and LS transistors. Smaller die area results in less switching power losses in spite of the increased Q_{xx} values.

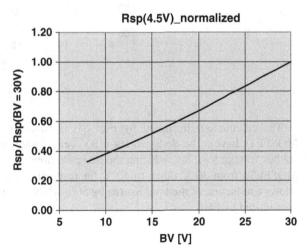

Fig. 6.3 Reduction of Rsp by low BV design of MOSFETs

Fig. 6.4 Impact of low voltage design on Q_{gd} and Q_g

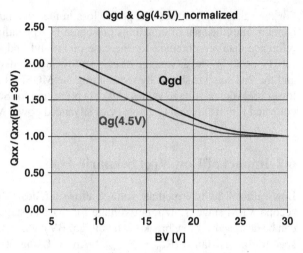

Fig. 6.5 Impact of low voltage design on Q_{oss} and Q_{rr}

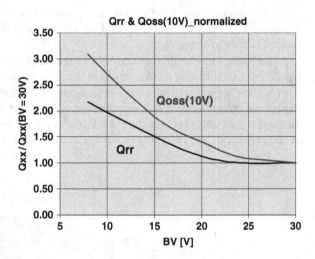

The calculation illustrated by Fig. 6.6 is based on the assumption that the MOSFETs have to be designed for a BV being twice the value of the intermediate bus voltage V_{IN}. So, reducing the designed breakdown voltage of the 2nd stage MOSFETs from 30 V down to 12 V for respective V_{in} values of 15 V and 6 V, allows a reduction of the total Si area by 50% which means that the MOSFET costs are reduced by half.

Fig. 6.6 Impact of low
voltage design on total power
loss and used Si area

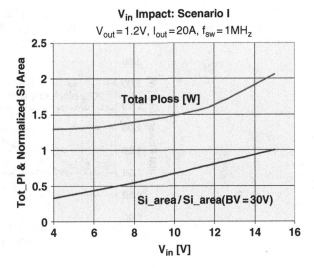

6.3 Impact of Low V_{IN}: Scenario II

The discussion of the 1st scenario has shown that a re-design of the MOSFETs used in the 2nd stage of a two stage sync buck converter for lower breakdown voltage can be used as an effective tool to reduce the area of the total Si used, i.e. to reduce the cost of the power switches. Here, in the different approach of the 2nd scenario, it is shown that the low BV re-design of the MOSFETs is a powerful tool in improving the performance of the converter and reduction of the total system cost.

In the 2nd scenario the total Si die area is kept the same. The partitioning of the size of Si dies for the HS and LS switches is adjusted according to the duty cycle. Lowering V_{in} changes the ratio of the output voltage to the input voltage which results in a longer time of conduction of the HS switch and less conduction loss in the LS switch. Accordingly, the die size ($R_{ds,on}$) of the HS switch should be increased and the die area of the LS switch should be reduced to minimize the total power loss. To this end following rule for die size partitioning is used:

$$LS\_area/HS\_area = SQRT((1 - DC)/DC)$$

Keeping total die size constant and lowering R_{sp} by low BV re-design leads to strong reduction of the conduction losses. The switching power loss decrease with lower V_{in} anyhow, as the power loss during switching is proportional to the switched voltage. The above discussion is illustrated by calculation results presented in Fig. 6.7.

Fig. 6.7 Impact of low
voltage design on total power
loss at high frequency

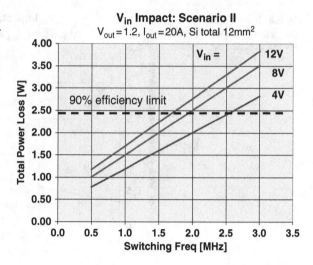

The reduction of the total power loss as shown in Fig. 6.7, allows increase of switching frequency of the converter while assuring the same high level of power conversion efficiency. As known, higher switching frequency in turn allows an improvement of the performance of the converter by making the input and output voltage regulation more effective. At the same time, higher switching frequency results in smaller size of the output filter what reduces the cost of the inverter.

Chapter Summary

Following topics have been handled in this chapter:

Section 6.1:

• Feasibility of high frequency switching

Section 6.2:

• Low BV re-design for MOSFET cost reduction

Section 6.3:

• Low BV re-design for improved performance and converter cost

Appendix: Power Loss Calculation

Current flow paths in a synchronous buck converter during on and off phases are illustrated in Fig. 1. It has to be noticed that following parameters are interrelated:

- the ratio of V_{OUT} to V_{IN} determines the required duty cycle (DC)
- R_{LOAD} determines the required output current I_{OUT} equal to I_{RMS}
- I_{RMS}, DC and output inductor L determine the ripple current described by low and high current peak values $I_{SW(on)}$ and $I_{SW(off)}$

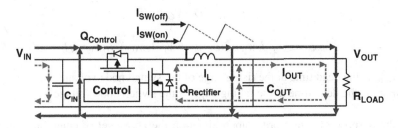

Fig. 1 Current flow in a synchronous buck converter

Power loss in the control switch:

$$P_{total} = P_{on-state} + P_{turn-on} + P_{gate} + P_{OSS}$$

$$P_{on-state} = I_{RMS}{}^{2*}R_{ds,on}$$

- $R_{ds,on}$ is compensated for VGS and temperature

$$P_{turn-on} = \frac{f^{*}V_{IN}{}^{*}I_{SW(on)}{}^{*}Q_{SW}{}^{*}R_{drive}}{2^{*}(V_{GS} - V_{th})}$$

- Q_{SW} is Q_{gd} plus Q_{gs} portion above Vth
- R_{drive} is R_g of the MOSFET plus the gate drive source impedance

J. Korec, *Low Voltage Power MOSFETs*, SpringerBriefs in Applied Sciences and Technology 7, DOI 10.1007/978-1-4419-9320-5, © Jacek Korec 2011

$$P_{turn-off} = \frac{f^* V_{IN}^* I_{SW(off)}^* Q_{SW}^* R_{drive}}{2^* V_{th}}$$

– R_{drive} is R_g of the MOSFET plus the gate drive sink impedance

$$P_{gate} = Q_g^* V_{GS}^* f$$

– Q_g is compensated for V_{GS}

$$P_{oss} = \frac{1}{2}^* Q_{OSS}^* V_{GS}^* f$$

Power loss due to rectifier switch:

$$P_{total} = P_{on-state} + P_{rr} + P_{gate} + P_{diode}$$

$$P_{on-state} = I_{RMS}^{2*} R_{ds,on}$$

$$P_{rr} = f^* Q_{rr}^* V_{IN}$$

– Q_{rr} includes Q_{rr} and Q_{oss} when they are separated in the data sheet

$$P_{gate} = Q_g^* V_{GS}^* f$$

$$P_{diode} = (V_f^* I_{SW(on)}^* t_{DT(on)} + V_f^* I_{SW(off)}^* t_{DT(off)})/f$$

– $t_{DT(on)}$ is the dead time at the rising edge of the switch node voltage
– $t_{DT(off)}$ is the dead time at the falling edge of the switch node voltage

For the purpose of a first order approximation the above set of equations can be further simplified by substituting $I_{SW(on)}$ and $I_{SW(off)}$ current values by I_{RMS}, as it has been done for calculations presented in the chapter on high frequency switching.

Closing Remarks

The material presented in this booklet is a mix of basic, public domain information, and of previously unpublished opinions based on 25+ year in-field expertise of the author. In both cases there is no good reference literature to the single statements as collected in this publication. This is the reason why the author decided to cite a few reference books covering power devices and applications in a general way instead of putting together a long list of technical papers and technical notes.

Still, it is author's expectation that this booklet will fill the existing gap in written publications in terms of offering a qualitative understanding of power MOSFET physics and design impact on their performance in power management applications.

References

1. S.M. Sze, "Semiconductor Devices, Physics and Technology", Wiley, New York, NY, 2001, 2nd Ed.
2. B.J. Baliga, "Fundamentals of Power Semiconductor Devices", Springer, New York, NY, 2008.
3. N. Mohan, T.M. Undeland and W.P. Robbins, "Power Electronics: Converters, Applications, and Design", Wiley, New York, NY, 1995, 2nd Ed.
4. Y. Bai, "Optimization of Power MOSFETs for High Frequency Synchronous Buck Converter", PhD Dissertation in Electrical Engineering, Virginia Polytechnic Institute, Aug. 2003.

Subject Index